PROFESSIONAL STANDARD OF THE PEOPLE'S REPUBLIC OF CHINA

Standard for Design of Railway Buildings

TB 10011-2012

Chief Development Organization: The Third Railway Survey and Design Institute Group Corporation
Approval Department: Ministry of Railways of the People's Republic of China
Implementation Date: March 19, 2012

China Railway Publishing House

Beijing 2016

图书在版编目(CIP)数据

铁路房屋建筑设计标准:TB 10011-2012:英文/中国铁路总公司组织编译. —北京:中国铁道出版社,2016.5
ISBN 978-7-113-21519-4

Ⅰ.①铁… Ⅱ.①中… Ⅲ.①铁路—交通运输建筑—建筑设计—行业标准—中国—英文 Ⅳ.①TU248.1-65

中国版本图书馆CIP数据核字(2016)第037118号

Chinese version first published in the People's Republic of China in 2012
English version first published in the People's Republic of China in 2016
by China Railway Publishing House
No. 8, You'anmen West Street, Xicheng District
Beijing, 100054
www. tdpress. com

Printed in China by Beijing Jinghua Hucais Printing Co., Ltd.

© 2012 by Ministry of Railways of the People's Republic of China

All rights reserved. No part of this publication may be reproduced or transmitted in any form or by any means, electronic or mechanical, including photocopying, recording, or by any information storage and retrieval systems, without the prior written consent of the publisher.

This book is sold subject to the condition that it shall not, by way of trade or otherwise, be lent, resold, hired out or otherwise circulated without the publisher's prior consent in any form of binding or cover other than that in which it is published and without a similar condition including this condition being imposed on the subsequent purchaser.

ISBN 978-7-113-21519-4

Introduction to the English Version

This version, as one of China Railway Corporation standards in English series, is translated under the organization of China Railway Economic and Planning Research Institute authorized by China Railway Corporation as per relevant procedures and regulations.

This standard is the official English language version of *Standard for Design of Railway Buildings* (TB 10011-2012) published by China Railway Publishing House. China Railway Corporation owns the copyright of this English version. According to *the Circular on Request for Opinions on Management Interface of Existing Railway Engineering Standards and Standards in Compilation* (Document Guo Tie Zong Ke Fa Han [2014] No. 176) issued by the General Affairs Department of National Railway Administration, the Chinese version of TB 10011-2012 has been put under the management of China Railway Corporation.

The English version was prepared by The Third Railway Survey and Design Institute Group Corporation. In case of discrepancies between the Chinese version and the English version, the former shall prevail.

Your comments are invited for revision of this version and should be addressed to China Railway Economic and Planning Research Institute, No. 29B, Beifengwo Road, Haidian District, Beijing, 100038, P. R. China.

The translation was performed by Zhao Jianhua, Zhang Jialu, Zhang Baoqun, Lu Jing, Wen Jizhou, Yang Jing, Du Junmei, Geng Yidi, Huang Jianshuang, Li Xin, Wei Qiang, Yang Weihong, Cui Xiulong, Chen Kun, Geng Liansong, Lin Mingjia, Xin Yu, Xiong Fei.

The translation was reviewed by Chen Shibai, Yang Sibo, Wang Lei, Gan Jiandong, Liang Jin, Song Huaijin, Liu Yanchao.

Notice of China Railway Corporation on the Issuance of the English Version of 16 Railway Engineering Standards including *Technical Specification for Geological Prediction of Railway Tunnel*

Document Tie Zong Jian She [2016] No. 94

The English version of the following 16 railway engineering standards is hereby issued: *Technical Specification for Geological Prediction of Railway Tunnel* (Q/CR 9217-2015), *Technical Specification for Construction of High-speed Railway Subgrade* (Q/CR 9602-2015), *Technical Specification for Construction of High-speed Railway Bridge and Culvert* (Q/CR 9603-2015), *Technical Specification for Construction of High-speed Railway Tunnel* (Q/CR 9604-2015), *Technical Specification for Construction of High-speed Railway Communication System* (Q/CR 9606-2015), *Technical Specification for Construction of High-speed Railway Signaling System* (Q/CR 9607-2015), *Technical Specification for Construction of High-speed Railway Electric Power System* (Q/CR 9608-2015), *Technical Specification for Construction of High-speed Railway Traction Power Supply System* (Q/CR 9609-2015), *Technical Code for Risk Management of Railway Construction Project* (Q/CR 9006-2014), *Interim Provision for Engineering Design of Natural Disaster and Foreign Object Intrusion Monitoring System for Railway* (Document Tie Zong Jian She [2013] No. 86), *Technical Specification for Construction of Railway Water Supply and Sewerage Works* (Q/CR 9221-2015), *Construction and Acceptance Standard of Railway Natural Disaster and Foreign Object Intrusion Monitoring System* (Q/CR 9745-2014), *Standard for Laboratory Management of Railway Construction Project* (Q/CR 9204-2015), *Guide for the Construction of Green Corridor for Railways* (Document Tie Zong Jian She [2013] No. 94), *Standard for Design of Railway Buildings* (TB 10011-2012) and *Technical Specification of Falsework for Cast-in-place Construction of Railway Concrete Girders* (TB 10110-2011). In case of discrepancies between the English version and the Chinese version, the latter shall prevail.

China Railway Economic and Planning Research Institute and China Railway Publishing House are authorized to publish these standards.

China Railway Corporation
April 12, 2016

Notice on the Issuance of
Standard for Design of Railway Buildings

Document Tie Jian She [2012] No. 51

Standard for Design of Railway Buildings (TB 10011-2012) is hereby issued and comes into effect on the date of issuance. *Standard for Design of Railway Buildings* (TB 10011-98) and *Interim Provisions for Design of Railway Equipment Housing* (Document Tie Jian She [2010] No. 63) issued by the Ministry of Railways are withdrawn.

Construction Management Department of Ministry of Railways is responsible for the explanation of this Standard, and Institute of Railway Engineering Technical Standards and China Railway Publishing House are authorized to publish this Standard.

Ministry of Railways of the People's Republic of China
March 19, 2012

Foreword

This Standard is developed on the basis of revising the *Standard for Design of Railway Buildings* (TB 10011-98) as per the requirements of *Notice on Issuing the Development Plan of Railway Engineering Construction Standards, Quotas and Standard Designs in* 2003 (Document Tie Jian She Han [2003] No. 41) and *Notice on Issuing the Development Plan of Railway Engineering Construction Standards in* 2012 (Document Tie Jian She Han [2012] No. 88) issued by the Ministry of Railways.

The setting principles and design requirements of railway buildings used for railway production, office work, vocational education and auxiliary facilities are determined in this Standard based on summarizing recent years' achievements in railway productivity layout adjustment and according to the requirements of railway construction development.

This Standard consists of 4 chapters: General Provisions, Basic Requirements, Production Buildings, and Auxiliary Buildings. Five appendices are also included.

The main revisions are as follows:

1. In Chapter 1 "General Provisions", the design principles of unified planning, moderate centralization and comprehensive arrangement are added.

2. Chapter 2 "Site Selection and General Layout" of the previous edition is revised as "Basic Requirements" of this Standard; some requirements for railway production buildings are added including seismic fortification, seismic design, fire protection, flood control, lightning protection, electromagnetic compatibility and earthing system.

3. Chapter 3 "Production and Office Buildings" of the previous edition is revised as "Production Buildings" of this Standard; moreover, some contents are added or revised as follows:

(1) Adding the principles of scale determination of production buildings, as well as the common requirements of design for the buildings such as section communication base station, repeater, signal relay station, and traction substation.

(2) Adding the design indices of building areas for passenger transport buildings.

(3) Adjusting the usable areas of freight business hall buildings in comprehensive freight yard according to the volume of freight transport.

(4) Adding the requirements on arranging the buildings used for EMU, dispatching, HVAC and firefighting equipment.

(5) Adding the requirements on the composition and configuration of the frontier (port) station buildings, information buildings and disaster monitoring buildings; deleting the satellite ground station, microwave station and repeater station.

(6) With regard to railway maintenance buildings, adding the buildings used for production dispatching, data analysis, metering and detection, comprehensive machine maintenance, etc.

(7) With regard to water supply buildings, adding the buildings used for passenger train water supply, sewage discharge central control room, drinking water treatment room, etc.

(8) With regard to railway public security buildings, adding the requirements for buildings in

police service areas and police boxes for railways with a speed of 250 km/h and above.

(9) Adjusting the area indices of the buildings used for staff vocational education and train crew apartments.

(10) With regard to health care and epidemic prevention house, adding the waste transfer stations and plateau area oxygen supply rooms.

4. In Chapter 4, "Auxiliary Buildings" and "Domestic Buildings" of the previous edition are merged into "Auxiliary Buildings" of this Standard, and the following revisions are made:

(1) Adjusting the building area indices of staff bathrooms.

(2) Deleting the staff health station and living supplies station along the railway.

(3) Deleting the staff residence, guesthouse, local bathroom, primary and secondary schools, nursery and kindergarten, medical treatment and public health center, and cultural activity center.

(4) Adjusting the area indices for staff dormitory.

The provisions printed in bold type are compulsory and must be enforced strictly.

All relevant organizations are kindly requested to collect data and summarize experience in the implementation of this Standard. In the case that any corrections or supplements are needed, please send the suggestions to The Third Railway Survey and Design Institute Group Corporation (No. 10, Zhongshan Road, Hebei District, Tianjin, 300142), and copy to Economic and Planning Research Institute of the Ministry of Railways (No. 29B, Beifengwo Road, Haidian District, Beijing, 100038) for future revisions.

Construction Management Department of the Ministry of Railways is responsible for the explanation of this standard.

Chief Development Organization:

The Third Railway Survey and Design Institute Group Corporation.

Participating Development Organization:

Economic and Planning Research Institute of the Ministry of Railways.

Chief Drafters:

Zhao Jianhua, Zhang Jialu, Sun Hongfeng, Zhou Xiaogang, Chen Jun, Shen Liewen, Wang Qiang, Wang Yinchuan, Wang Shu, Cheng Yali, Yang Weihong, Ji Huimei, Zhang Zhifang, Zhang Dongming, Yang Bingfeng, Han Zhao, Li Xiaobing, Zhang Guoliang, Chen Jie, Mou Zhongxia, Wang Guoliang, Xia Tianyan, Qi Yana, Diao Pengzhi, Du Junmei.

Contents

1 General Provisions ·· 1
2 Basic Requirements ·· 2
  2.1 General Requirements ·· 2
  2.2 Site Selection ·· 3
  2.3 General Layout ··· 3
3 Production Buildings ·· 5
  3.1 General Requirements ·· 5
  3.2 Buildings for Passenger Transportation ·· 6
  3.3 Buildings for Loading/Unloading and Storage of Freights ····················· 6
  3.4 Operating Rooms in Station ·· 8
  3.5 Border (Port) Station Building ·· 10
  3.6 Rooms for Systems of Communication, Signaling, Information,
      Disaster Prevention and Safety Monitoring ·· 10
  3.7 Railway Maintenance Buildings ·· 13
  3.8 Buildings for Traction Power Supply and Power Distribution System ······· 15
  3.9 Locomotive Maintenance Buildings ·· 18
  3.10 Rolling Stock Maintenance Buildings ·· 20
  3.11 EMU Buildings ·· 21
  3.12 Water Supply and Drainage Buildings ·· 21
  3.13 Dispatching Station Buildings ·· 23
  3.14 Equipment Rooms for Heating, Ventilation and Air Conditioning (HVAC) and
       Fire-fighting ··· 24
  3.15 Railway Real Estate Management Buildings ·· 24
  3.16 Railway Public Security Buildings ··· 24
  3.17 Office and Vocational Training Rooms ··· 25
  3.18 Crew Apartment ·· 26
  3.19 Hygiene and Epidemic Prevention Room ··· 27
4 Auxiliary Buildings ··· 28
  4.1 Staff Canteen ·· 28
  4.2 Staff Bathroom ··· 28
  4.3 Staff Dormitory ··· 29
Appendix A Building Area of Rooms in Railway Workshop, Work Sub-section, EMU
            Servicing Depot and Work Section ·· 30
Appendix B Technical Requirements of Communication Equipment Rooms ················· 31
Appendix C Technical Requirements for Signaling Equipment Room ···················· 35
Appendix D Technical Requirements for Information, Disaster Prevention and Safety

Monitoring Equipment Rooms	36
Appendix E　Technical Requirements of Traction Power Supply and Power Distribution Buildings	38
Explanation of Wording in this Standard	40
Citations	41

1 General Provisions

1.0.1 This standard is prepared to meet the needs of railway transportation and production, and enable the railway building design to comply with the basic requirements of safety, adaptability, economy and sanitation.

1.0.2 This standard is applicable to the building design of new railway projects and railway reconstruction projects.

1.0.3 Railway buildings shall be designed based on the principle of economical land use and less or no cultivated land occupation, and shall comply with the requirements of the state on energy conservation and environmental protection.

1.0.4 The railway buildings shall be planned uniformly, centralized moderately and arranged comprehensively.

1.0.5 The construction land for the railway buildings shall be planned based on long-term design period. The criteria and scale of building construction shall be determined according to the planning of railway transportation development, local economy, climate and other local conditions.

1.0.6 The new technology and new materials which are mature, reliable, cost-effective, safe and adaptable shall be employed in the design of railway buildings.

1.0.7 Railway reconstruction project should be designed by making full use of existing buildings and facilities.

1.0.8 The design of railway buildings shall not only comply with the requirements stipulated in this Standard, but also comply with those stipulated in the relevant current standards of the state.

2 Basic Requirements

2.1 General Requirements

2.1.1 The Seismic fortification category of the following buildings shall be Class B: the buildings used for traffic dispatching, operation, communication, signal, power supply and water supply of high-speed railway, passenger dedicated railway, Class I & II main lines of passenger and freight mixed railway, and freight dedicated railway, as well as the waiting hall buildings of large or extra-large railway station with the maximum number of passengers gathered in waiting hall being 6 000 and above.

2.1.2 Seismic design, fire-proof design and structural safety design of railway buildings must comply with the relevant requirements stipulated in current standards of the state.

2.1.3 The design of equipment buildings in station area shall comply with the following requirements:

　　1　The equipment buildings used for communication, signaling, information and disaster monitoring, as well as the substation and distribution substation with voltage below 10 kV, the distribution room, refrigeration machine room, air condition machine room, etc. shall be built jointly with the railway station building. If there is no possibility for them to be built jointly with the railway station building, they should be built jointly with other railway buildings.

　　2　Buildings shall be partitioned according to their functions and shall be relatively centralized. Buildings which have a close relation to each other shall be arranged adjacently.

　　3　The area of equipment buildings without extension conditions should be determined based on the operating requirements of equipment in the long term, and these buildings shall be comprehensively employed in the short term.

　　4　The rooms with heavy equipment or where the equipment produces loud noise or intensive vibration should be arranged on the ground floor where the influence on other rooms is relatively small.

　　5　All kinds of pipelines shall be designed and planned uniformly. An integrated pipeline trench should be installed for pipelines of the same route.

　　6　For unmanned equipment building, exterior window should not be arranged, and safety door shall be provided.

2.1.4 Building design shall comply with the requirements of equipment installation and maintenance.

2.1.5 Deformation joint of building shall not be placed in the equipment machine room.

2.1.6 With regard to railway equipment buildings, the lightning protection, electromagnetic compatibility, earthing, vibration control, dust prevention, electro-static discharge (ESD) protection, damp proofing, rat proofing, etc. shall comply with the relevant requirements stipulated in current standards of the state.

2.1.7 When going through the partition wall or floor slab of equipment building, the electric cable, cable shaft, cable trench (groove), etc. shall be provided with fireproof sealing.

2.1.8 The building decoration shall be in accordance with the requirements of *Code for Fire Prevention in Design of Interior Decoration of Buildings* (GB 50222), and environment friendly materials shall be used. The floors and walls with dust prevention and ESD protection requirements shall be smooth and neat.

2.1.9 Measures against water overflow and leakage shall be taken when water-consumption facilities are provided in equipment buildings.

2.1.10 Dedicated stand-alone air conditioners shall be provided in the important equipment rooms which are used for communication, signaling, information, disaster monitoring, etc. The configuration of air conditioners shall meet the requirements of regular operation during maintenance period.

2.2 Site Selection

2.2.1 Site selection for railway buildings shall comply with the following requirements:

1 The flat area with high elevation, effective drainage, favorable development prospect and convenient traffic condition should be selected.

2 The area with serious unfavorable geological conditions such as debris flow, landslide, karst and fault developing zone shall not be selected.

3 The site shall be located far away from the areas where severe pollutant is generated such as hazardous substance and a large amount of dust/soot, or where flammable, explosive or radioactive substance is stored.

4 **The site shall not be located within the regions influenced by high voltage cable corridor or by significant underground project and underground pipes.**

5 Existing buildings should not be demolished on a large scale.

2.2.2 Design of equipment buildings in the range of railway with the speed of 200 km/h and above shall comply with the following requirements:

1 Elevation of outdoor ground surface shall comply with the relevant requirements of flood level and waterlogging level stipulated in current national standards. The return period of flood level and waterlogging level shall comply with the requirements in Table 2.2.2.

Table 2.2.2 Return Period of Design Flood Level or Waterlogging Level

Name of building	Return period of design flood level and waterlogging level (years)
Traction substation	100
Coupling post, switching post, autotransformer post, electric substation, power distribution substation, section communication & signal facility building; pump station of water supply station	50
Pump station of domestic water supply station (point)	20

Notes: 1 The design flood level or waterlogging level for outdoor ground surface of equipment buildings shall be increased by 0.5 m for safety.
    2 If the pump station is located on the bank of great river, lake and reservoir, wave run-up shall be added to the outdoor ground elevation.
    3 If production facility buildings used for communication, signaling, electricity, etc. are located at the same place, their foundation elevations shall be identical and the higher design flood level or waterlogging level shall be adopted.

2 Traffic condition shall be taken into consideration during site selection.

2.2.3 For production buildings and technical operation buildings closely related to railway line, their site selection may be conducted on the basis of operational needs and must comply with the relevant requirements of railway structure clearance and current standards of the state.

2.3 General Layout

2.3.1 General layout shall be planned uniformly. The layout of buildings other than those which must be arranged adjacent to railway lines to fulfill the needs of technical operation shall comply with the following requirements:

    1 General layout shall be in coordination with the plan of urban development.

    2 Function division shall be definite and layout shall be compact and reasonable, satisfying the requirements of both the production process and the reserved long-term land use of buildings.

    3 General layout shall comply with the national requirements on fire protection, occupational safety and health, environmental protection, energy saving and greening.

    4 Ground elevation shall be determined by considering the topographic and hydrologic conditions. For production buildings related to route elevation, the vertical design shall be conducted by considering the elevations of track surface, subgrade surface, flood level, and station platform surface.

2.3.2 Buildings shall be divided into production zone and auxiliary zone according to function division, and be provided with traffic roads.

3 Production Buildings

3.1 General Requirements

3.1.1 The area of production buildings and auxiliary buildings shall be determined by the scale of production, producing technique and requirements of facilities.

3.1.2 The scale of railway buildings shall comply with the following requirements:

1 The design of common office buildings, production buildings and auxiliary buildings shall be determined by short-term design period.

2 The design of inspection garages of locomotive, rolling stock and EMU shall be determined by short-term design period, reserving the condition for long-term design period.

3 The design of the following production buildings shall be determined by long-term design period:

    1) Passenger station buildings.

    2) Electrical equipment buildings of traction substation, electric substation and distribution substation.

    3) Signal building, dispatching station, and dispatching building of station.

    4) Buildings for communication and information equipment.

    5) Buildings for disaster monitoring equipment.

3.1.3 The production buildings and office buildings with similar functions located in the same district should be jointly built through unified planning. For new railways, a common repairing building should be set for the stations and depots of the same district.

3.1.4 Standby equipment shall not be provided with dedicated building. However, space (which shall have light partition walls and should serve as office or lounge at ordinary times) shall be available for using the standby equipment. For the above mentioned space, the position of cable holes, power sockets, integrated pipelines, indoor reserved terminals for equipment integrated grounding parts and building load shall be deployed as per the requirements of production equipment buildings.

3.1.5 Solid enclosing walls and anti-theft gates shall be set for communication base stations, repeaters, signal repeaters, traction substations, autotransformer stations, substations, distribution substations, switching post, and coupling post. The height of the wall shall not be less than 2.5 m. Metal net or caltrop with height not less than 0.7 m shall be set on the wall, and the upper part of the net with width of 20 cm shall incline outwards with the angle of inclination being 45 degree.

3.1.6 The water supply and drainage pipes and the water consuming equipment shall not be set for rooms for communication, signaling, information and disaster monitoring as well as the adjacent rooms above them, otherwise water-proof measures shall be taken, which shall comply with the related requirements in the current *Code for Design of Electronic Information System Room* (GB 50174).

3.1.7 The standards of offices and vocational training room in stations and depots shall comply with the requirements in Section 3.17. The standards of rooms in workshops (work subsection and EMU station) and work areas shall comply with the requirements in Appendix A.

3.1.8 Garages (sheds) may be set in stations, depots, workshops (work subsection and EMU station) and work sections in accordance with the needs of operation, land use and architectural design.

3.1.9 The safety distance between oil depot and other buildings shall be in accordance with the requirements of the current standard *Code for Fire Protection Design of Buildings* (GB 50016) and *Code for Design on Fire Prevention of Railway Engineering* (TB 10063).

3.2 Buildings for Passenger Transportation

3.2.1 The buildings for passenger transportation mainly include passenger transportation buildings in station and those in passenger transport depot. The former shall comply with the requirements of the current standard *Code for Design of Railway Passenger Station Buildings* (GB 50226) and other relevant standards, and the latter shall mainly include the following rooms (facilities):

    1 Preparing rooms, waiting rooms and retiring rooms for train crew.

    2 Management office, conductor's rooms, appointing rooms, work shift rooms (study rooms) and dressing rooms.

    3 Bill room and cash receiving room.

    4 Spare beddings rooms, special transportation spare parts rooms, rooms for miscellaneous materials and supply rooms.

    5 Food processing rooms and commodity rooms (including refrigeration rooms).

    6 Rooms for cleaning and sorting of beddings.

    7 Serving rooms such as spare parts rooms and quality inspection rooms.

3.2.2 Preparing rooms for train crew shall be set for passenger transport depots, whose usable floor area shall be determined on the basis of the maximum number of train crew on duty according to the standard of 1.0 m^2 per person, and should not be less than 30 m^2.

3.2.3 The usable floor area of retiring rooms in passenger transport depot shall be determined based on the actual crew number according to the standard of 4.0 m^2 per person.

3.2.4 The usable floor area of management offices (conductor's room included) of each group of train crew should be 60 m^2. The usable floor area of work shift rooms (study rooms) should be determined based on the number of the largest group of train crew according to the standard of 1.0 m^2 per person.

3.2.5 The usable floor area of dressing rooms of each group of train crew should be determined based on the number of crew according to the standard of 0.45 m^2 per person.

3.2.6 The building area of beddings room should be determined based on the number of operating sleeping cars according to the standard of 25 m^2 per car. The building area of rooms for miscellaneous materials and commodity rooms (including refrigeration rooms) should be determined based on the actual conditions of the project.

3.2.7 The building area of food processing rooms and washing rooms for beddings shall be determined based on the workload and techniques.

3.2.8 The building area of service rooms such as spare parts rooms and quality inspection rooms shall be determined based on the actual needs.

3.3 Buildings for Loading/Unloading and Storage of Freights

3.3.1 Buildings for freight transportation shall be mainly composed of the following rooms:

    1 Office, business hall, workshop.

    2 Study (roll call) room, room for outdoor duty staff.

    3 Truck scale inspection room, integrated inspection and monitor room.

3.3.2 The design of offices and business rooms for freight transportation shall comply with the following requirements:

1 The offices and business rooms should be jointly built.

2 The building shall be located at the place convenient for the cargo owners to transact business.

3 For Large and medium freight yards, the business office and business hall should be separated by low counters. The business office of small freight yards should be enclosed.

4 The building area of freight business hall in comprehensive freight yards should be determined according to Table 3.3.2. The building area of freight business hall in freight yards with capacity equal to or larger than 3 000 000 tons shall be determined based on actual conditions and should not exceed 500 m².

Table 3.3.2 Building Area of Freight Business Hall (Office) in Comprehensive Freight Yards

Scale of freight yard	Small	Medium	Large	
Weight of received and dispatched freights (million tons)	below 30	31~100	101~199	200~299
Usable area (m²)	80~100	101~200	201~300	301~400

3.3.3 The usable area of roll call/study room shall be determined based on the number of the largest group of train crew with the standard of 1.0 m² per person.

3.3.4 The room for outdoor duty staff should be located at appropriate place in the warehouse area or the end of ordinary warehouse, and it's building area should be 15 m². Duty stuff room and other rooms shall not be arranged in the warehouse.

3.3.5 Truck scale inspection room should be located at the entrance of freight yard. Comprehensive inspection and surveillance rooms and offices for train transportation shall be jointly built.

3.3.6 Loading/unloading building shall be mainly composed of the following rooms:

1 Office for Loading/unloading, roll call/study room, retiring room for loaders, tools and materials room.

2 Maintenance room for loading/unloading equipment, forklift truck maintenance room, storehouse for loading/unloading equipment, pallet repairing room, fuel storage.

3 Room for maintenance of loading/unloading equipment.

4 Container loading/unloading warehouse and related platform.

5 Container maintenance workshop and maintenance shed.

6 Container loading/unloading warehouse.

3.3.7 The design of loading/unloading buildings shall comply with the following requirements:

1 The offices and lounges shall be jointly built.

2 The number of beds in the retiring rooms shall be determined by the number of the largest group of loaders, the useful area of retiring room should be 4.0 m² per bed for loading/unloading drivers and 3.0 m² per bed for other loaders.

3 The usable area of tools and materials room shall be determined based on the quantity of tools.

4 The roll call (study) rooms shall be designed in accordance with Article 3.3.3.

3.3.8 The building area of loading/unloading equipment maintenance rooms shall be determined according to the maintenance cycle and the quantity of equipment.

3.3.9 Besides the current related national standards, dangerous and explosive goods warehouses shall also comply with the following requirements:

1 Different kinds of dangerous goods in the same warehouse shall be stored in separate rooms.

2 Dangerous goods warehouses shall be kept dry and well-ventilated. The roof shall be heat-insulated and the windows shall be adumbral.

3 The floor and dado of dangerous goods warehouses shall be acid/alkaline-resistant and convenient for brushing and drainage.

4 Washing facilities shall be set around the dangerous goods warehouses.

5 The explosive goods warehouses shall be single-storey building. The usable area of each warehouse shall not begreater than 100 m² and the headroom should not exceed 3.6 m.

3.3.10 The design of radioactive goods warehouses shall comply with the following requirements:

1 Underground construction may be used and reinforced concrete structure shall be used for the related walls and roofs.

2 The doors and windows shall be prevented from exposure to radioactive rays. The indoor wall surface shall be flat and smooth and the wall corners shall be arc-shaped.

3 The ground shall be convenient for washing and floor drains shall be set on the ground. Natural ventilation facilities shall be set at the roof.

3.3.11 The design of fuel warehouse shall be in accordance with the requirements of *Code for Design and Construction of Filling Station* (GB 50156). The safety distance between the fuel warehouse and other buildings shall be in accordance with the requirements of *Code for Fire Protection Design of Buildings* (GB 50016) and *Code for Design on Fire Prevention of Railway Engineering* (TB 10063).

3.3.12 Buildings of loading/unloading maintenance center shall comply with the requirements of Article 3.1.3. When the maintenance center is set up separately, it should be built together with loading/unloading maintenance buildings in larger freight yards. The building area shall be determined according to the maintenance cycle and the quantity of loading and unloading equipment.

3.3.13 The bathroom of stations handling dangerous goods shall be provided with shower rooms dedicated for the staff contacting dangerous goods.

3.4 Operating Rooms in Station

3.4.1 The operating rooms in station mainly include the following:

1 Station duty room, driving room for outdoor staff, switchman's cabin, switch cleaner's cabin, station crossing guard room.

2 Station dispatching (freight transport dispatching) building, car number checker room, shunting warden room, hump centralized control building, hump yard top operator (lift hook operator) room, air-pulling operator room, hump tail centralized control building, shunting team resting room, braking operator (connect operator) room, cast brake shoe storage room, hump tail protecting room, sluice building.

3 Train-tail attendant room, train-tail operator room, train-tail machinery compartment, battery charging room.

4 Calling room of workshop (duty shift room, study room), dressing room, lounge, equipment tools storage room, maintenance room.

3.4.2 If a station has centralized signal building, the station duty room shall be set in the signal building. Monitor engineer room shall be set if a station adopts centralized traffic control (CTC) system. Duty room shall be arranged separately for other stations.

3.4.3 The driving room for outdoor staff should be located at the side of the middle or end of the arrival and departure track, and it shall face the station yard facilities. It may also be constructed together with other houses if the operation requirements are met.

3.4.4 Switchman's cabin should be located near or with an effective distance from the busy turnout.

3.4.5 Switch cleaner's cabin should be situated at the switch-dense location of electrically centralized interlocking station, and it may also be combined with other buildings nearby.

3.4.6 The guard room at level crossing in station shall be located at the right side of the railway line.

3.4.7 The station dispatching (freight dispatching) building should be located at the centre of station, and it may be combined with signal cabin, station office building and train operation room, etc. The station dispatching building shall face the track and be endowed with outlook in three directions. The station dispatching controller room, station duty office, general duty room, station dispatching room, working schedule director room, and station freight dispatching room shall be integrated into a dispatching hall, otherwise the station dispatching room, freight dispatching room and car number checker room (indoor) shall be arranged adjacently.

3.4.8 Car number checker room (outwork) shall be set in the train yard with receiving and departure operation. The car number checker room should be located at the middle or the end of the receiving and departure track, and it may be combined with other buildings.

3.4.9 For stations with hump, tail centralized control building should be set at one side of the track in the switch area of hump tail; for stations without hump, dispatching warden room should be set at throat area of marshalling line group or outside the throat area of hump tail marshalling line group. Dispatching warden room should be combined with dispatching resting room, car number checker room and switchman's cabin.

3.4.10 Hump centralized control building should be located under the hump in order to be convenient for watching the hump dispatching operation and the left side of rolling direction.

3.4.11 Hump top operator (hook lifting operator) room should be located at the left side of the hump platform in the pushing direction.

3.4.12 The air-pulling operator room should be located outside the middle of the yard with receiving and departure operation, and it may be combined with car number checker room.

3.4.13 Dispatching group lounge shall be set in the station with dispatching stuff, and it should be combined with dispatching warden room, car number checker room and switchman's cabin.

3.4.14 Braking operator (connect operator) room shall be combined with cast brake shoe storage room, and it should be located between the middle and the tail of the dispatching yard.

3.4.15 The hump tail protective room shall be located at the tail of the dispatching yard.

3.4.16 Sluice building should be located at the starting/ending point of the throat area between station and locomotive depot/EMU depot.

3.4.17 The train tail duty room, train tail operator room, train tail machinery compartment, battery charging room and storage room should be set at the tail of the yard train, and they may be built together considering specific conditions.

3.4.18 In medium and small station, the charging chamber should be set near the duty room; in marshalling station, the charging chamber may be set centrally or separately according to different car yards, and the charging chamber should be located at the place where shunting operators are relatively concentrated. Charging chamber should have two rooms, one for charging, another for maintenance equipment and spare tools.

3.4.19 The roll call room (duty shift room, study room), dressing room, lounge, equipment tools storage room and maintenance room should be arranged based on station level and the number of users.

3.5 Border (Port) Station Building

3.5.1 National frontier port buildings mainly include buildings used for passenger and freight check-in, customs service, inspection for border goods, flora and fauna, and health quarantine.

3.5.2 Buildings for freight transport and loading/unloading shall comply with related requirements of Section 3.3.

3.5.3 Joint inspection (customs service, inspection for border goods, flora and fauna, and health qarantine) rooms of passenger station buildings shall be in accordance with the current *Code for Design of Railway Passenger Station Buildings* (GB 50226).

3.5.4 Joint inspection building for freight transport shall comply with the requirements of joint inspection of state customs.

3.6 Rooms for Systems of Communication, Signaling, Information, Disaster Prevention and Safety Monitoring

3.6.1 Rooms for communication system mainly include communication equipment rooms and communication inspection & repair rooms.

3.6.2 The communication equipment rooms may be set at the station building, signaling equipment room, signaling relay station in sections, base stations in section, repeater, traction substation (TS), switching post (SP), sectioning post, auto-transformer post (ATP), substation & distribution substation, and other premises where communication facilities are required, such as communication section, maintenance workshop, maintenance work area and independent station.

Communication equipment rooms are mainly comprised of communications machine room, power supply room, battery room, network management room, duty room, lounge, monitoring room, cable lead-in room, spare parts and instrument room, spare electric generator cabin, oil storage and spare room, and an integrated room may be set as required based on actual need.

3.6.3 The communication inspection rooms may be set up at the communication section, communication workshop and communication work section.

Communication inspection rooms are mainly comprised of dispatching and commanding, operation monitoring and data analysis rooms (including network management, data management, technical support, security center and monitoring room), metering room (including instrument testing and maintenance), emergency repair equipment warehouse, material warehouse (shed), etc.

Communication workshop includes field workshop, conference dispatching room, centralized repair room (medium scale repair team), wireless maintenance point and communication maintenance point. The inspection rooms of communication workshop are mainly comprised of duty office, monitoring room, technical support room, testing room, inspection room and warehouse (shed).

Communication work section includes field work section, wireless base inspection section, train broadcast section, etc. The inspection rooms of communication work section mainly include duty room, operating room, locker room and warehouse.

3.6.4 Telegragh cable post shall be provided with acceptance area, work area, locker room, lounge and warehouse. The acceptance area should not be less than 10 m^2. The size of work area, lounge and warehouse should be determined according to the needs of work. The average seating work area in the phone cable post should not be less than 4 m^2.

3.6.5 The area and layout of the communication equipment room should be in accordance with the following requirements:

    1 The area of the communication equipment room may be determined by the actual needs of the project taking into consideration the short and long term development.

    2 The communication equipment used for substations of electric power system and those used for traction power supply system shall be arranged separately. If it is not allowed, they may be arranged in the control room, and the work area of these two systems shall be relatively isolated.

    3 The communication machine room of the intermediate station, overtaking station and block house shall be close to the station operation room, signaling machine room, passenger traffic machine room and power distribution room.

    4 The layout of communication equipment room shall reserve the conditions for the access of antenna and feeder of wireless communication system, and the conditions for the leading-in of optical and electrical cables with different routes.

    5 Communication equipment rooms should be arranged in quiet places, far from the electromagnetic interference source and far from the places susceptible to strong vibration.

    6 When communication equipment rooms are integrated with offices, separation measures should be taken.

    7 The indoor temperature and humidity shall be in accordance with the requirements in Appendix B.

3.6.6 Communications equipment rooms shall conform to the relevant requirements in *Specifications of Engineering Design for Telecommunication Private Premise* (YD/T 5003), *Classification Standard for Seismic Protection of Telecommunication Buildings* (YD 5054), and *Code for Design of Electronic Information System Room* (GB 50174). When the container type computer room is used in section, the relevant railway industry standards shall be conformed to.

3.6.7 Communication equipment rooms should be integrated with communication inspection room and other production rooms.

3.6.8 The buildings for signaling system shall mainly consist of the following:

    1 The buildings used for signaling and communication division shall mainly include dispatching command and data analysis rooms for signaling and communication system (including information management system equipment for signaling and communication system, centralized supervision & monitoring system for signaling equipment, onboard data analysis, etc.), metering room (including instrument and meter test and maintenance), test room for signaling and communication division, material warehouse, etc.

    2 The buildings used for signaling maintenance shall mainly include signaling workshop, special mission workshop, signaling section, maintenance and test section, etc. The special mission workshops include signaling maintenance workshop, medium scale repair workshop, electronic equipment workshop, onboard equipment workshop (including automatic train protection (ATP), cab signal, train operation monitoring and recording device (LKJ), etc.) and hump workshop.

    3 The signaling equipment buildings are composed of station signaling equipment room and cable room, signaling equipment room or container equipment room for relay stations, central equipment room and equipment monitoring room for centralized traffic control system/train dispatching and commanding system (CTC/TDCS), radio block center (RBC) room and equipment monitoring room, equipment room for level crossing signaling system, equipment room (including computer room) for hump signaling system, hump control equipment room, etc. The station signaling equipment room shall consist of signaling computer room, relay room and power supply room, lightning-proof cable frame room.

3.6.9 Appendix A and the following requirements shall be met for the buildings of signaling workshop and work section:

1 Analysis room (the decoration shall comply with the standards of computer room), dispatching room, and warehouse (shed) shall be allocated in signaling workshop, among which the warehouse shall have the function of storing emergency spare parts.

2 Emergency duty office and spare parts room shall be allocated in the station without signaling section.

3.6.10 The location and layout of signaling equipment rooms shall comply with the following requirements:

1 The distance from outer edge of signaling equipment room to the centre of the nearest railway track shall not be less than 7.00 m.

2 The independent entrances and exits shall be configured when signaling equipment room is combined with other rooms.

3 When signaling equipment room is combined with station buildings, cable room shall be allocated on the ground floor at the side close to the railway line and shall be connected with outdoor cable trough. Cable frame or holes for signaling cables shall be reserved between cable room and equipment room.

4 Signaling equipment room and operation room shall be arranged adjacently.

5 The area of the signaling building shall be determined based on such factors as the scale of station or yard, type of interlocking and blocking system, configuration requirements of signaling equipment, location and plan of future development, etc.

3.6.11 The following requirements on signaling equipment room shall be complied with:

1 The clear distance of the passage for the transportation of signaling equipment in the building shall not be less than 1.50 m.

2 Windows may not be provided or few windows may be provided for signaling equipment room and windows should not be provided for unmanned signaling equipment room. Bi-layer airtight windows or double-glass airtight windows shall be provided and sunscreen measures shall be taken if windows are necessary. When airtight windows are provided or no windows are provided, ventilation facilities shall be installed.

3 The lightning-proof, electromagnetic compatibility, earthing, quakeproof, anti-dust, ESD protection, moisture-proof, fire-proof, and rodent-proof, etc. of signaling equipment room shall meet the requirements of the current *Code for Design of Electronic Information System Room* (GB 50174) and the current relevant railway industry standards of lightning-proof, electromagnetic compatibility and earthing.

4 The signaling equipment room shall comply with not only the technical requirements on equipment layout, equipment operation, maintenance and inspection, but also the technical requirements in Appendix C.

5 Container type equipment rooms for signaling system used in section shall comply with the relevant railway industry standards.

3.6.12 The signaling workshop and signaling section should be allocated in signaling building or combined with other system rooms or office.

3.6.13 The information rooms mainly include information equipment room, information repair room, duty room and spare parts room. The information equipment rooms include computer room of information, integrated supervision room, fire-fighting control room, electrical and mechanical control room, power supply room, information wiring and device room.

3.6.14 Information room may be set in station building, depot (house) and other places with information system.

3.6.15 The layout plan of information equipment room shall comply with the following requirements:

1 The fire-fighting control room and electrical and mechanical equipment control room in station building should be combined.

2 Integrated supervision room and computer room of information should be adjacent to the communication equipment room, and the computer room for the public security information management system shall be arranged next to the police office room.

3 Entrance and exit of information equipment room shall meet the transport requirements of information equipment.

4 Fire-fighting control room should be set either at the outer wall of first floor, or at the basement, with the emergency exit leading to the outside.

3.6.16 Information equipment room shall comply with the following technical requirements:

1 The clear width of the room shall not be less than 3.3 m, information wiring rooms on different floors shall be provided with a cable shaft.

2 Independent door shall be set for the information wiring room.

3 The equipment arrangement of information equipment room shall comply with the requirements of computer room management, operation and safety, equipment and material transport, equipment installation, operation and repair, and inspection.

4 The technical requirements of the information equipment room shall comply with the requirements of Appendix D.

3.6.17 Disaster prevention and safety monitoring rooms mainly include disaster prevention and safety monitoring computer room, duty room, repair room and spare parts room. Disaster prevention and safety monitoring computer room may be set in the dispatching office, station, base station, relay station of signaling system, traction substation, autotransformer station, sectioning post, etc.

3.6.18 Disaster prevention and safety monitoring computer room shall be provided with an independent outdoor, and device arrangement in the computer room shall comply with the requirements of Item 3 of Article 3.6.16.

3.6.19 Disaster prevention and safety monitoring repair room shall be set in the depot, workshop, and work section according to the work division of equipment management. Repair rooms for depot and workshop mainly include duty room, supervision room and technical support room.

3.7 Railway Maintenance Buildings

3.7.1 The railway maintenance buildings mainly include the following:

1 Track maintenance division (bridge maintenance division) buildings mainly include dispatching room (production scheduling control center), inspection and control center (data analysis center), metering room (calibration and testing center), archives room, material storage (shed), railcar shed, oil storage and old track material storage.

2 Besides meeting the requirements of Article A.0.1, workshop buildings of passenger and freight mixed railway shall be provided with oil storage and material storage (shed).

Besides workshop buildings of passenger and freight mixed railway, the machinery room, repairing room, instrument room and railcar shed shall also be set in workshop buildings of high-speed railway and passenger dedicated railway.

3 Besides meeting the requirements of Article A.0.2, workshop buildings of passenger and freight mixed railway shall also be provided with track patrolling room, oil storage, machine room, material storage (shed) and repairing room.

Besides production buildings in passenger and freight mixed railway, railcar shed and other equipment maintenance rooms shall also be set in high-speed railway and passenger dedicated railway maintenance buildings.

4 Rush repair post and materials room of passenger dedicated railway.

5 Duty room and rush repair materials room of block post.

6 Station and line maintenance workshop and track maintenance section buildings of junction station, station and line track maintenance section buildings of EMU depot.

7 Supervision room, tunnel ventilation equipment room.

8 Buildings of track and mechanical division, overhaul division, as well as infrastructure maintenance base of passenger dedicated railway may be set as needed.

3.7.2 The buildings of track maintenance division, workshop and track maintenance section shall comply with the following requirements:

1 The buildings of track maintenance division (bridge maintenance division), workshop and track maintenance section shall be comprehensively planned and grouped into office area, inspection area, power area, material storage area, etc.

2 The data analysis center, metering room and instrument room of track maintenance division, workshop and track maintenance section shall be jointly built with office buildings.

3 Metering room and instrument room shall be kept away from sources of vibration and electromagnetic field.

3.7.3 Station track maintenance workshop and track maintenance section buildings shall be set in junction station. Station and line track maintenance section buildings shall be set in EMU depot.

3.7.4 High-speed railway and passenger dedicated railway stations without workshops and maintenance section shall be provided with rush repair rooms, which shall be combined with other maintenance rooms.

3.7.5 Block post should be provided with supervision room and rush repair materials room.

3.7.6 The area of grade crossing watching post with one person on duty should be 18 $m^2 \sim$ 20 m^2. The area of grade crossing watching post with two persons on duty should be 21 $m^2 \sim$ 25 m^2. Grade crossing watching room shall have good lookout condition.

3.7.7 The design of railcar shed shall comply with the following requirements:

1 The length of railcar shed shall comply with the requirements of parking and maintenance of two motor cars and two platform lorries.

2 The ground in front of track garage door shall be hardened, its elevation shall be in consistent with the pavement elevation and rail surface elevation in track garage.

3 The length of the inspection pit in the track car garage shall comply with the requirements of inspection.

3.7.8 The length and span of material shed and the lifting machine configuration shall comply with the requirements of the storage, loading/unloading of common rail and turnout.

3.7.9 The area of comprehensive machine repair workshop shall be determined according to the requirements of machinery and process.

3.7.10 Buildings of large-scale track maintenance machinery division, overhaul division and rail welding base, if needed, shall be set separately.

3.8 Buildings for Traction Power Supply and Power Distribution System

3.8.1 Buildings of traction power supply and electric power distribution system are mainly composed of the following:

1 Maintenance workshop of power supply section.

2 Insulating oil storage, oil storage room and OCS maintenance vehicles garage of power supply section.

3 Power supply workshop, electric power maintenance workshop, OCS maintenance workshop, power distribution maintenance workshop, as well as garages of ladder trolley cars and OCS maintenance vehicles.

4 Traction substation, switching post, sectioning post, auto-transformer post, power substation, power distribution as well as rooms for backup diesel generator, power distribution, control, pipeline.

3.8.2 Buildings of power supply workshop shall be combined with OCS maintenance workshop.

3.8.3 Buildings of OCS and electric power maintenance workshop may be set separately in large stations such as junction station and district station. Buildings of maintenance workshop may be set near large stations or the station where the substation is included. It may also be combined with power supply workshop and electric power workshop.

3.8.4 Buildings of OCS maintenance workshop may be composed of ladder trolley car garage, maintenance room, material storeroom, tools' storeroom, rush repair material storeroom, OCS maintenance vehicle garage, etc.

3.8.5 The design of OCS maintenance vehicle garage shall comply with the requirements of maintaining and parking. OCS maintenance vehicle component maintaining workshop shall be set according to maintenance technical requirements.

The length of OCS maintenance vehicle garage shall comply with the requirements of rail vehicle parking. Shift room, tools storeroom, materials storeroom, diesel wareroom and toilet shall be planned as a whole when the OCS maintenance vehicle garage is jointly built with OCS maintenance workshop. Shift room, tools storeroom, materials storeroom, diesel wareroom and toilet shall be provided if the OCS maintenance vehicle garage and the OCS maintenance workshop are built separately.

3.8.6 The design of electric power overhaul workshop shall comply with the following requirements:

1 The ceiling height of the high voltage test cubicle shall comply with the technological requirements. Security isolation facilities for observation shall be set between the operating area and the high-voltage area.

2 The floor of the transformer maintenance room and the oil treatment room shall be corrosion-resistant. The ground level of the oil treatment room should be lower than the outdoor terrace. The interior natural ventilation shall be favorable. The ceiling height shall comply with the requirements of transformer hoisting, hanging core and maintenance. The transformer store room or shed shall be set in transformer maintenance area. The instrument room and tools room may be set in the electric power overhaul workshop.

3 Besides being kept far away from vibration equipment, the instrument room, relay room and microcomputer-based protection control room shall have dustproof, anti-vibration, damp-proof, ventilation, and good lighting.

3.8.7 Insulated oil storage design shall comply with the following requirements:

1 When adopting above-ground horizontal type oil tank, the clear height of oil depot shall not be less than 4.8 m. The fire resistance class shall not be lower than Class 2.

2 Indoor floor shall be provided with anti-oil-flow facilities.

3 Steel blind window and metal net should be set in ventilation-hole outer wall. The height of ventilation hole shall not be less than 0.3 m.

4 When shed-style buildings are adopted, the oil storage shall be provided with solid walls at least 2.5 m high.

3.8.8 The design of traction power supply equipment building shall comply with the current *Design Code of Railway Electric Traction Feeding* (TB 10009) and the following requirements:

1 Traction power supply equipment buildings should consist of high-voltage room or high-voltage gas insulated switchgear (GIS) room, cable interlayer, control room, capacitor room, transformer room, maintenance room, spare parts and tools room, etc.

2 The area and layout of traction power supply equipment building shall be compact and reasonable.

3 Fire prevention design of capacitor building equipped with combustible medium shall comply with the relevant requirements of the current *Code for Fire Protection Design of Buildings* (GB 50016) and *Code for Design on Fire Prevention of Railway Engineering* (TB 10063).

4 Transformer room shall be set separately when the total oil weight of indoor oil immersed transformer exceeds 100 kg.

Oil blockage or oil storage facilities shall be set when the total oil weight of indoor single electrical equipment exceeds 100 kg. The volume of oil blockage should be determined as 20 percent of the total amount of oil. Emergency oil shall be discharged to safe place. When oil blockage facilities cannot be set, oil storage facilities shall be set to accommodate all the oil.

5 Solid materials should be adopted for power distribution wall. The indoor passageway shall not be passed through by irrelevant pipelines.

6 Lifting devices shall be set in high-pressure gas insulated switchgear (GIS) room, and the lifting capacity shall comply with the lifting requirements of the heaviest equipment.

7 Traction power supply equipment buildings shall comply with equipment layout, equipment operation and maintenance, overhaul requirements, as well as the requirements of Appendix E.

3.8.9 Duty room and bathroom shall be set in traction substation and switching post, and its area should not be less than 20 m^2 per building.

3.8.10 The layout of power equipment buildings shall comply with the following requirements:

1 Cable interlayer should be set in independent multi-layer 10 kV distribution substation, 35 kV and above substation. The height of mezzanine shall not exceed 2.2 m. The width of main road shall not be less than 3.5 m, and shall have conditions for U-turn of vehicles.

2 The high calorific equipment and high-voltage capacitor should be placed in different rooms from other equipment.

3 The heavy current, high voltage and high-rate impact load power supply equipment buildings should not adjoin the electronic information shielding equipment buildings which are easily interfered with by the electromagnet; otherwise, protection measures shall be taken.

3.8.11 The design of power equipment buildings shall comply with the following requirements:

1 When adopting oil-immersed transformer, oil reservoir shall be set in the substation attached to the main building.

2 A three-phase transformer or voltage regulator whose oil content is over 100 kg shall be isolated in an explosion-proof room which is equipped with fire control facilities.

3 In the power distribution equipment room, fireproof materials shall be used to seal the inlet and outlet of the cable, as well as any void around it.

4 In lightning-proof design of the power equipment buildings, measures shall be taken to prevent direct lightning and lightning surge invasion. The metal shell of roof equipment, the metal sheath of cables and the building metal components shall be earthed.

5 When the large-scale power equipment is installed on the second floor or above, a lifting platform shall be set, and its structure and size shall comply with the requirements of hoisting the largest equipment.

6 The rooms equipped with computer and microcomputer-based protection devices shall comply with requirements on humidity, temperature and dust proof. Emergency lighting shall be installed in the control room, at the location of the indoor power distribution equipment, as well as in the main passages.

3.8.12 The design of power equipment buildings shall comply with the following technical requirements:

1 The distance between the bare bus on the top of the high-voltage switchgear and the indoor crown plate or the beam bottom shall not be less than 900 mm; for the non-bare bus it shall not be less than 300 mm.

2 For the indoor oil-immersed transformers which require in-situ maintenance, the ceiling height may be determined by the minimal lifting height of the transformer core plus 700 mm, and the indoor width may be determined by the width of two sides plus 800 mm.

3 When the length of low-voltage distribution panel (including the panel board) is over 6 m (5 m for high-voltage distribution panel), two exits leading to the room or other rooms shall be set behind their panels. When the distance between the two exits of low-voltage distribution panel is over 15 m, more exits shall be set.

4 When the length of the distribution room is over 7 m, two exits shall be set at the two ends of it. When the distribution room is arranged on the second floor and the floors above, there shall be a door leading to the outdoor safety evacuation staircases. The number of control room entrance shall not be less than two. Two entrances should be set in the cable interlayer.

5 The distance between the windowsill of the power distribution room and the outdoor ground should not be less than 1.8 m. Windows should not be set at the side of the power distribution room facing the street. Control room shall have good lighting and ventilation conditions. Measures shall be taken to prevent the rain, snow and small animals from entering the transformation and distribution rooms.

6 Interior decoration design of the power equipment building shall comply with the following requirements:

1) Cement or terrazzo floor should be adopted in the transformer room and capacitor room.
2) The interior walls and floor which are easy to clean and maintain shall be adopted in control room, power distribution room, relay room, auxiliary rooms and other rooms requiring high cleanness.

7 The heating ventilation and air conditioning design for the power equipment building shall comply with the following requirements:

1) The indoor temperature of the control room should be 16℃~18℃ in winter and not more than 30℃ in summer.
2) The transformer room and the capacitor room shall be provided with ventilation

condition. The indoor temperature shall not be higher than 40℃ in the capacitor room, relay room and distribution room, and shall not be higher than 45℃ in the oil-immersed transformer room.

    3) Mechanical ventilation shall be applied in the rooms with gas-insulated equipment (GIS), and the air inlet shall be located at the bottom of the room.

3.9 Locomotive Maintenance Buildings

3.9.1 Locomotive maintenance buildings mainly include the following:

1 Production buildings for locomotive operation, servicing, inspection, etc; train crew apartment and crew waiting room, rescue facilities buildings.

2 Locomotive inspection and repair workshop, test workshop and auxiliary room.

3 Locomotive auxiliary inspection and repair room, locomotive monitoring facilities maintenance room, equipment maintenance room, dynamic facilities housing, material warehouse (shed) and spare parts warehouse.

4 Locomotive dispatching room, information computer room, technical room and section acceptance room.

5 Production buildings for locomotive turnaround depot, locomotive turnaround point and locomotive transference point.

6 Locomotive fuel storage and fuel unloading/delivering room.

3.9.2 The arrangement of locomotive maintenance buildings shall be in compliance with the following requirements:

1 The overall planning shall be made for the arrangement of buildings and facilities. Locomotive operation, servicing, repair and maintenance, auxiliary buildings, office, domestic buildings, etc. shall be arranged according to their functions.

2 Production buildings generating harmful substances shall be arranged on the weather side of the local annual minimum frequency wind.

3 Workshop producing strong vibration and high noise shall be arranged separately at places with less impact on the environment, and shall be in accordance with the requirements of the current *Code for Design of Noise Control of Industrial Enterprises* (GBJ 87).

4 Locomotive fuel storage shall be arranged separately, and the safety distance from other buildings shall be in accordance with the requirements of the current *Code for Design of Oil Depot* (GB 50074) and *Code for Design on Fire Prevention of Railway Engineering* (TB 10063).

3.9.3 The design of locomotive inspection and repair workshop shall be in accordance with the following requirements:

1 The shape of inspection and repair workshop shall be simple and neat, and the common components up to national standards should be adopted.

2 When the inspection and repair workshop is located in section with unfavorable geological conditions, strengthening measures shall be taken, and the operating requirements of monolithic ballast bed and inspection pit shall be met.

3 For the inspection and repair workshop with crane, the design height of lower chord of roof truss shall meet the spatial requirements of crane equipment installation, maintenance and repair.

4 Inspection and repair workshop should take advantage of the natural lighting, and the lighting level should be Grade Ⅲ~Ⅳ. Roof lighting belt shall also be arranged besides the lighting windows on the lateral.

5 The building foundation and underground structure of the inspection and repair workshop shall be coordinated with the arrangement of the equipment foundation and inspection pit, meeting the requirements of installing equipment foundation and pipeline.

6 The indoor floor elevation of the inspection and repair workshop should be consistent with the top of rail, and should be higher than the outdoor ground by 0.15 m~0.30 m.

7 The material of ground finishes of the inspection and repair workshop and auxiliary room shall be determined according to the inspection and maintenance technology and cleaning requirements.

8 Drainage slope and water collection facilities shall be used on the indoor ground floor in inspection and repair workshop and auxiliary room.

9 Production buildings generating harmful substances shall be equipped with separate ventilation facilities.

10 For the inspection and repair workshop, the entrance shall be provided with porch and canopy, and heat insulation door shall be adopted in the severe cold region. The ground in front of the entrance shall be hardened, and the size of the entrance opening shall comply with the requirements of the current standard *Structure Gauge for Standard Gauge Railways* (GB 146.2).

11 Crew dressing room and lounge shall be arranged in the inspection and repair workshop and auxiliary room, and restrooms shall be arranged at the proper place according to the arrangement of production buildings.

3.9.4 The design of locomotive servicing shed and casual faults handling shed shall be in accordance with the following requirements:

1 The longitudinal outside end length of the shed shall be 3 m larger than the outside end of the inspection pit when locomotive inspection pit is set in locomotive servicing shed and casual faults handling shed. Smoke exhaust facilities shall be installed on the shed roof in cases of servicing and casual faults handling for diesel locomotive.

2 Shading and rain proofing structures should be provided outside the locomotive servicing shed and casual faults handling shed.

3 When the electrified overhead contact line system is lead into the locomotive servicing shed or casual faults handling shed, electrostatic induced earthing devices shall be set in the building structure of the shed. The clear height of the shed shall be in compliance with the requirements of the installation height and safety distance of OCS.

3.9.5 For the production rooms such as measuring room, assay room, instrument room and bearing room that have special requirements on cleanness, temperature and humidity, airtight door and window shall be provided and interior decoration shall be designed according to the requirements for inspection and repair.

3.9.6 Production buildings with acid/alkali corrosion and oil pollution shall be in accordance with the requirements of the current *Code for Anticorrosion Design of Industrial Constructions* (GB 50046).

3.9.7 The arrangement of locomotive painting workshop and painting drying room shall be in accordance with the requirements of the current *Code for Design on Fire Prevention of Railway Engineering* (TB 10063), and dust control and ventilation facilities shall be installed in the workshop.

3.9.8 The design of locomotive load test station and locomotive complete test workshop shall be in accordance with the following requirements:

1 Doors and windows shall be provided at the corresponding place of testing locomotive in the operation control room, and soundproof measures shall be taken.

2 When the noise level in the workshop after treatment cannot meet the requirements of the current *Code for Design of Noise Control of Industrial Enterprises* (GBJ 87), noise reduction workshop shall be adopted.

3.9.9 In addition to being in accordance with the requirements of Section 3.18, train crew apartment for locomotive depot shall include train crew study room and safety monitoring room.

3.9.10 The design of locomotive maintenance buildings shall be in accordance with the requirements of the current *Code for Design of Railway Locomotive Facilities* (TB 10004).

3.10 Rolling Stock Maintenance Buildings

3.10.1 Rolling stock maintenance buildings mainly include the following:

1 Repairing shed, bogie room, wheel axle depot, antifriction bearing room, hook buffer cabin, brake room, spare parts inspection and maintenance center and other maintenance workshops of passenger and freight car depot.

2 Equipment maintenance workshop, power facilities, parts and material warehouse and garage of passenger and freight car depot.

3 Passenger train technical servicing point.

4 Station repair yard.

5 Passenger train inspection and service point.

6 Freight train inspection and service point.

7 Tank washing point.

8 Buildings for train operation safety monitoring system and automatic train identification facilities.

3.10.2 Overall planning shall be made for the arrangement of buildings in the passenger and freight car depot, and the buildings shall be grouped into different areas according to their functions such as train inspection and auxiliary production, power and equipment, materials storage, office and residence.

3.10.3 Production buildings such as repairing shed and maintenance workshop shall be designed according to the requirements of technology.

3.10.4 Besides Items 1, 4, and 10 of Article 3.9.3, the repairing shed for car depot shall also be in accordance with the following requirements:

1 The clear height of door openings shall not be less than 5.5 m in ice-cooled refrigerator car workshop and tank car shed.

2 The elevation of floor of the repair shed shall be lower than the top of rail by 10 mm.

3 Drainage slope and water collection facilities shall be used on the indoor floor in passenger car repairing shed and servicing workshop.

3.10.5 The floor of repairing shed shall be higher than the outdoor ground level by 0.15 m to 0.30 m, and the slope of ramp way shall meet the requirements for access of battery forklift.

3.10.6 When the repairing shed and maintenance workshop are located in section with unfavorable geological conditions, strengthening measures shall be taken.

3.10.7 Crew dressing room and lounge shall be arranged in the repairing shed and maintenance workshop, and restrooms shall be arranged in the proper place according to the arrangement of production buildings.

3.10.8 The design of repairing shed shall be in accordance with the requirements of Item 2 of Article 3.9.4, and the indoor floor elevation shall be lower than the top of rail by 10 mm.

3.10.9 The building foundation and underground beam of the repairing workshop (shed) and inspection workshop shall be coordinated with the arrangement of the pipelines such as the drainage pipes of equipment foundation pit and inspection pit.

3.10.10 The design of production workshop with crane shall be in accordance with Item 3 of Article 3.9.3.

3.10.11 The production workshops in car depot, such as brake room, antifriction bearing room, isothermal assembling room, oil pressure shock-absorber inspection and repair shed, measuring room, detecting room, and chemical laboratory, that have special requirements on cleanness, temperature and humidity shall be in accordance with Article 3.9.5.

3.10.12 Buildings with acid/alkali corrosion and oil pollution shall be in accordance with Article 3.9.6.

3.10.13 The production workshop generating poisonous and hazardous gas and dust shall be in accordance with Item 2 of Article 3.9.2 and Item 9 of Article 3.9.3.

3.10.14 Painting workshop and paint shop in car depot shall be in accordance with Article 3.9.7.

3.11 EMU Buildings

3.11.1 EMU buildings mainly include the following:

1 Servicing, inspection and casual faults handling production building for EMU, operation duty cab for EMU driver, crew apartment, crew room and lounge, crew apartment and crew room for EMU mechanician.

2 Maintenance workshop and auxiliary workshop for EMU.

3 Equipment maintenance workshop, power equipment building, material warehouse (shed), spare parts warehouse, and flammable goods warehouse for EMU depot.

4 Dispatching center, information computer room, technical room and section acceptance room.

5 Production buildings for EMU servicing depot and stabling yard.

6 Signal tower for EMU depot (running shed).

7 Ice melting and snow removing workshop and washing workshop may be set as needed in severe cold regions.

3.11.2 The arrangement of EMU buildings shall be in accordance with Items 1~3 of Article 3.9.2.

3.11.3 The design of inspection workshop, wheel lathe house and maintenance workshop shall be in accordance with Article 3.9.3.

3.11.4 The design of production buildings for measuring, assay, instrument, bearing, etc. shall be in accordance with Article 3.9.5.

3.11.5 The design of the building with acid/alkali corrosion and oil pollution shall be in accordance with Article 3.9.6.

3.11.6 Crew apartment for EMU driver and EMU mechanician shall be in accordance with Article 3.9.9.

3.11.7 The design of EMU production buildings shall be in accordance with the current standards for design of high-speed railways.

3.12 Water Supply and Drainage Buildings

3.12.1 Water supply and drainage buildings mainly include water supply plant (station), water supply pressure station, fire-fighting pump cabin, centralized monitoring room for water supply of passenger train, waste discharge centre monitoring room, resting room for passenger train water supply and waste discharge, drinkable water treatment station for station building, sewage treatment plant (station), vacuum station (centre), drainage pump room, etc.

3.12.2 Water supply and drainage buildings shall be in accordance with the following requirements:

1 Facilities for heating, ventilation, lighting and drainage shall be installed according to production requirements in equipment room, which shall meet the technical requirements on moisture proof, fireproof, noise prevention, and vibration reduction.

2 Disinfection room, chemical-dosing room and chemicals storage room which generate poisonous and hazardous gas shall not be located at underground or semi-underground places with poor ventilation conditions, and safety measures shall be taken for doors and windows.

3.12.3 The arrangement of water supply and drainage buildings shall be in accordance with the following requirements:

1 The arrangement of water supply plant (station) and sewage treatment plant (station) shall meet the requirements of equipment transport and installation.

2 Lounge for passenger train water supply and waste discharge, centralized monitoring room, and drinkable water treatment station shall be combined with the station building.

3 Drainage (wastewater) pump station should be built separately, and ventilation measures shall be taken.

3.12.4 Equipment room for water supply and drainage shall meet the following requirements:

1 Equipment room doors shall accommodate the largest equipment. The clear height of the room shall meet the requirements of equipment installation, operation and inspection.

2 The minimum clear height of rooms in pump station shall be in accordance with Table 3.12.4.

Table 3.12.4 **Minimum Clear Height of Rooms in Pump Station**

Type of room		Clear height (m)
Pump house	Without lifting equipment	3.5
	With lifting equipment	4.0
Diesel generator room		4.0~4.5
Tube well pump house		4.5
Medicine storage room and disinfector chamber		4.0
Auxiliary room		3.2
Vacuum waste discharge center		4.0

3 The duty room of attended pump house shall be adjacent to pump room. Soundproof watching windows and soundproof doors shall be installed for the partition wall.

4 Lifting hole shall be arranged on the roof of tube well pump house. The dimensions of the lifting hole shall be the outer dimensions of lifting devices plus 0.2 m on each side for safety.

5 Noise reduction measures shall be taken in pump room, blower room, and centrifugal dehydrating equipment room. Noise control of machinery room shall meet the requirements of the current *Code for Design of Noise Control of Industrial Enterprises* (GBJ 87).

6 Chlorination room shall be provided with one door that opens outwards leading directly to the outside and one fixed observation window, and it must be partitioned from other workrooms. Anti-corrosion measures shall be taken for the floor and wall in chlorination room.

7 Ventilation facilities shall be installed in chlorine storeroom, disinfector chamber (chlorination room) and chemicals storage room. Distribution facilities shall meet the anti-corrosion and explosion poof requirements.

8 Chemical laboratories shall have good ventilation and lighting conditions, the materials of wall, floor and operation console shall be corrosion resistant and easy to clean. The sterile room and chemicals storage of chemical laboratory should be located on the shady side.

9 Technical requirements for water supply and drainage monitoring room shall be in comply with the requirements for device room in Appendix D.

3.12.5 In addition to being in accordance with the requirements of Article A.0.2, water supply and drainage section buildings shall have inspection and repair room.

3.13 Dispatching Station Buildings

3.13.1 Dispatching station buildings mainly include the following:

1 Rooms for signaling, communication, information and disaster prevention and safety monitoring, traction power supply and electric room, as well as rooms for train dispatching, passenger dispatching, freight dispatching, planning dispatching.

2 The signal rooms of dispatching station shall include central equipment room for centralized traffic control system/train dispatching and commanding system (CTC/TDCS), work area maintenance center for centralized traffic control system/train dispatching and commanding system (CTC/TDCS), power supply room, network management center room. The work area rooms include duty room, operation room, changing room, spare parts room. Network management center room shall meet the standards for design of equipment room. The work section rooms of dispatching section shall meet the standards for design of signaling work area rooms.

3 Communication rooms for dispatching station mainly include equipment rooms and maintenance rooms. The equipment rooms include communication equipment room, network management room, duty room, monitoring center and spare parts room; maintenance rooms include communication work section duty room, operation room, dressing room and spare parts room.

4 The information rooms and disaster prevention and safety monitoring rooms of dispatching station include equipment room, maintenance center, network management room, duty room, spare parts room, etc. The equipment rooms mainly include information computer room, fire control room, power supply room, information wiring and equipment room, large-screen control room.

5 Traction power supply and electric rooms for dispatching station include dispatching room, remote control equipment room, power supply room and auxiliary room. The area of dispatching room shall be determined by the quantity of dispatching console and the configuration of remote control equipment.

3.13.2 The area and other requirements of dispatching station buildings shall be in accordance with the following requirements:

1 A combined dispatching hall shall be built for all the dispatching buildings for train, passenger transport, freight transport, planning, etc.

2 The design of train dispatching buildings shall be based on operating distance, density of operating trains, number of stations, traffic capacity, lead-in number of special railway lines and branch lines, and corresponding dispatching console.

3 The passenger traffic dispatching building shall be designed on the basis of the dispatching console allocated according to the passenger train operation within the administrative area.

4 The freight traffic dispatching building shall be designed on the basis of the dispatching console allocated according to the marshalling of freight traffic station, the operating quantity of loading/unloading stations and the operation amount within the administrative area.

5 The planning dispatching building shall be designed on the basis of the dispatching console allocated according to the conditions of loading/unloading truck, evacuation, handover train at dividing line, small operation train and passenger train within the administrative area.

6 Technical requirements of signal equipment room, information and disaster prevention and safety monitoring room and traction power supply equipment room in dispatching station shall be in accordance with the requirements of Appendices C, D and E.

3.13.3 The dispatching station of high-speed railway shall be provided with liaison office for railway maintenance dispatchers.

3.13.4 The equipment room of dispatching station shall be up to the standard of A-level stipulated in *Code for Design of Electronic Information System Room* (GB 50174).

3.14 Equipment Rooms for Heating, Ventilation and Air Conditioning (HVAC) and Fire-fighting

3.14.1 The equipment rooms include boiler room, heat exchange station, refrigeration equipment room, control room, ventilation and air conditioning equipment room, gaseous fire extinguisher storage room, high level water tank room and fire-fighting water pump house.

3.14.2 The scale, plan layout and requirement of equipment room for HVAC and fire-fighting shall be in accordance with the current standards of the state.

3.15 Railway Real Estate Management Buildings

3.15.1 Railway real estate management buildings mainly include rooms for depot, workshop, work section and maintenance point, material processing and maintenance equipment room, and depot warehouse (shed).

3.15.2 Building area of warehouse (shed) may be determined as follows: ten thousand square meters of building maintenance area corresponding to $4\ m^2 \sim 5\ m^2$ building area.

3.15.3 Rooms for depot, workshop and work section should be combined with other railway production buildings of similar characteristics in the same area.

3.15.4 The building area of procession room or material room constructed in work section may be $120\ m^2$ to $130\ m^2$. The building area of material shed may be $40\ m^2$.

3.15.5 The usable area of maintenance point of high-speed railway, passenger dedicated railway, large stations and super-large stations should be $30\ m^2$.

3.15.6 In addition to being in accordance with Article A.0.2, railway real estate management buildings shall also include comprehensive inspection and repair room.

3.16 Railway Public Security Buildings

3.16.1 The composition and equipment level of railway public security building shall be synthetically determined in accordance with the current building standard for public security service and technology, as well as the distribution of police staff within the administrative area of the railway, the length of route, the passenger and freight transport volume and the security situation along the railway route.

3.16.2 Railway public security building shall be constructed near station where the traffic is convenient.

3.16.3 Railway public security building should be constructed independently. When constructing the basic level police service building in combination with the neighboring railway office building, the independent zone and entrance/exit shall be set, and the appearance of the police service building shall be in accordance with the provisions stipulated by the Ministry of Public Security.

3.16.4 The police substation buildings shall be in accordance with the following requirements:

1 Police substation shall be constructed in railway marshalling station, district station, as well as large freight and passenger station and EMU running depot within 50 km to 70 km along the railway.

2 Two or three stationary police rooms shall be arranged at the station and the EMU serving depot in the administrative range of the police substation.

3.16.5 Training field and car park shall be established in the police substation.

3.16.6 The usable area of police room in station and duty point in freight yard should be 30 m^2 to 50 m^2.

3.16.7 When constructing the police service building and police sentry box for the railway with a speed of 250 km/h and above, their location and number shall be reasonably determined in accordance with project demand and the local conditions and shall comply with the following requirements:

1 Police service building shall be constructed beside railway station, communication base station and relay station, and shall be constructed at intervals of 15 km to 20 km (except for stations with police substation); when being constructed at railway station, it shall be built jointly with the station building. The building area of police service space should be 100 m^2 to 120 m^2. The police service space should be provided with water supply and drainage equipment (including lavatory and bathroom), power supply equipment, public security management information system, public network telephone, public security dispatching telephone, GSM-R mobile phone, video monitor terminal, etc.

2 The police sentry box shall be constructed outside railway protection fence at the two ends of long bridge and tunnel, emergency evacuation corridor, crossover bridge, as well as the subgrade and roads close to villages. The police sentry box may be constructed each 2 km. The police sentry box shall be constructed with consideration of the overall planning of both the maintenance corridor and the bridge rescue and evacuation corridor. The building area of the police sentry box shall be 4 m^2. The police sentry box shall be built in the form of masonry structure with spire, windows in four orientations and safety door. Police sentry box located in severe cold region and cold region should be equipped with heating facilities.

3.17 Office and Vocational Training Rooms

3.17.1 The offices of depot and station shall mainly include management office, technology office, conference (teleconference) room, reference (archives) room, store room, etc.

3.17.2 The building area of offices shall be determined according to the design capacity specified in Table 3.17.2 and in accordance with the requirements of the current *Design Code for Office Building* (JGJ 67).

Table 3.17.2 Building Area of Depot and Station Office

Capacity (number of persons)	20~40	41~100	101~200	201 and above
Building area per person (m^2)	17.0~18.0	16.5~17.0	16.0~16.5	16.0

Notes: 1 Chemical laboratory and the production rooms adjacent to office building such as workshop office and crew room are not included.
    2 Engineering archives room is not included.

3.17.3 The technical training rooms of the stations not pertaining to train depot and those of the depots should include the teacher's office, classroom and operation room. The building area of technical training rooms shall be determined in accordance with Table 3.17.3 based on the total headcount of station (depot).

Table 3.17.3 Building Area of Technical Training Room

Categories	Locomotive depot, EMU depot	Car depot, power supply depot, track maintenance depot, communication and signaling depot	Train operation depot, passenger transportation depot, and Class II and above stations
Building area per person (m²)	1.20~1.25	0.80~0.85	0.70~0.75

3.17.4 Technical training rooms of different depots should be jointly built in districts with favorable conditions.

3.18 Crew Apartment

3.18.1 Locomotive crew apartments, EMU crew apartments and train crew apartments should be built as an integrated apartment at a quiet place convenient for on/off duty. The integrated apartment shall be divided into locomotive, EMU and train crew sections with separate management according to the requirements of production and transportation as well as number of crew of different capacities.

3.18.2 Crew apartment shall be mainly composed of living rooms, public activity rooms (including study room, recreation room, reading room, television room), canteen, office, equipment room, etc.

3.18.3 The crew apartment shall be designed in accordance with the following requirements:

1 The recreation room shall be kept away from the living room.

2 The train crew apartment shall be provided with restroom and lavatory.

3 Restrooms without natural ventilation shall be provided with ventilation duct or other ventilation facilities.

4 Corridor or passage should be built between living area and activity area.

3.18.4 The living room shall be designed in accordance with the following requirements:

1 Locomotive crew apartments and EMU crew apartments shall be designed as single rooms. The living room shall be provided with air-conditioner and restroom. The restroom shall be provided with shower, basin and hot water.

2 Train crew apartments shall be designed as double rooms. The living room shall be provided with air-conditioner and restroom. The restroom shall be provided with shower, basin and hot water.

3 Living rooms shall have good orientation and natural lighting. Living rooms facing west shall be provided with shading devices.

3.18.5 Canteen shall be provided for all apartments. The usable floor area of dining hall, kitchen and auxiliary rooms of canteen shall be in accordance with the following requirements:

1 Two beds of the locomotive crew apartment and the EMU crew apartment should correspond to one seat in the dining hall.

2 Three beds of the train crew apartment should correspond to one seat in the dining hall.

3 The usable floor area of the dining hall should be determined on the basis of 1.1 m² per seat. The restaurant to kitchen area ratio should be 1 : 1.

3.18.6 The management office shall be in accordance with the following requirements:

1 The duty room shall be arranged at the main entrance of the apartment.

2 Apartments with more than 100 beds shall be provided with a director office and a management office.

3 Apartments with 100 or fewer beds shall be provided with an integrated office.

4 Apartments should be provided with a cleaning room and a drying room, and those with more than 100 beds shall be provided with a beddings-arrangement room, a sewing room and an ironing

room.

5 The usable floor area of duty room, director office, integrated office and spare parts room should be 13 m² ~ 15 m².

3.18.7 The building area of locomotive crew apartment and EMU crew apartment with not more than 200 beds may be determined in accordance with Table 3.18.7-1. The building area of locomotive crew apartment and EMU crew apartment with more than 200 beds may be determined according to actual condition. The building area of train crew apartment with not more than 400 beds may be determined in accordance with Table 3.18.7-2. The building area of train crew apartment with more than 400 beds may be determined according to actual condition. The building area of integrated apartment may be determined according to the number of locomotive crew, EMU crew and train crew.

Table 3.18.7-1 Building Area of Locomotive Crew Apartment and EMU Crew Apartment (m² per bed)

Type of Apartment \ Capacity	50 beds	100 beds	200 beds	300 beds	400 beds
Locomotive crew apartment, EMU crew apartment	27.5~29.0	26.5~28.0	25.5~27.0	—	—

Table 3.18.7-2 Building Area of Train Crew Apartment (m² per bed)

Type of Apartment \ Scale	50 beds	100 beds	200 beds	300 beds	400 beds
Train crew apartment	20.0~21.5	19.0~20.5	18.0~19.5	17.0~18.5	16.0~17.5

Note: The table does not include the building area of canteen, boiler room, air-conditioning room and distribution room.

3.19 Hygiene and Epidemic Prevention Room

3.19.1 The seat of railway administrations, provincial capitals, passenger transport terminals and freight transport terminals shall be provided with hygiene and epidemic prevention rooms, whose building area shall be determined in accordance with relevant standards of the state.

3.19.2 Large stations shall be provided with rooms for hygiene and epidemic prevention staff and articles, whose building area shall not be less than 60 m². Rooms dedicated to storing and preparing disinfector, insecticide and rat poison shall be provided with equipment for water supply and drainage, heating, and ventilation, and security measures shall be taken for the windows and doors of these rooms.

3.19.3 Stations with affiliated railway sanitation supervision agency shall be provided with station sanitation supervision room, whose building area shall not be less than 40 m².

3.19.4 For new railway lines on plateau, the station areas and the work areas along the line shall be provided with centralized oxygen room and medicine room, whose building area should be 40 m².

3.19.5 Stations and depots shall be provided with clinic in accordance with the scale of work area, professional hazards, and the number of workers.

3.19.6 Large railway passenger stations and EMU depots (posts) should be provided with refuse transfer station.

3.19.7 The site selection of refuse transfer station shall be in accordance with the urban master planning, in addition, the station shall be located at the weather side of the direction with minimum annual wind frequency and at the places with convenient transportation.

4 Auxiliary Buildings

4.1 Staff Canteen

4.1.1 Units with the maximum number of staff on duty not less than 240 may be provided with staff canteen. Units with the maximum number of staff on duty below 240 may be provided with dining house.

4.1.2 Canteen and dining house shall include dining room, kitchen and auxiliary room.

4.1.3 The capacity and building area of canteen and dining house shall be in accordance with the following requirements:

　　1　The number of seats in canteen shall be determined on the basis of 50 percent of the maximum number of staff on duty with consideration of stagger of dining time.

　　2　The building area shall be in accordance with Table 4.1.3.

Table 4.1.3　Building Area of Canteen and Dining House

Type of house	Canteen					Dining house
Capacity (number of seats)	60	100	200	300	400	
Building area (m² per seat)	3.4~3.6	2.8~3.1	2.5~2.7	2.4~2.5	2.3~2.4	2.6~3.4
Minimum usable floor area of dining room (m² per seat)	0.85~1.10					—

　　Notes: 1　When the maximum number of staff on duty is small, larger value shall be taken for the building area of dining house.
　　　　　2　When the canteen is provided with cooling room, its building area may be properly increased.

4.1.4 The building design of canteen shall be in accordance with the current *Code for Design of Restaurant Building* (JGJ 64).

4.2 Staff Bathroom

4.2.1 Basic-level production units with hygiene characteristics of Class I to Class IV shall be provided with staff bathroom or shower cubicle, which shall be designed in accordance with the current *Hygienic Standards for the Design of Industrial Enterprises* (GBZ 1).

4.2.2 Production units with the maximum number of staff on duty not less than 100 shall be provided with bathroom building; if less than 100, bathroom shall be provided.

4.2.3 The scale of staff bathroom of basic-level production units shall be in accordance with the following requirements:

　　1　The scale of bathroom shall be determined on the basis of 100% of the maximum number of staff on duty and one shower should meet the needs of 9 persons.

　　2　One shower shall correspond to three lockers. The ratio of men's lockers to women's lockers should be 8 : 2 for production workshop, 6 : 4 for office, and 5 : 5 for passenger transportation depot; the specific ratio shall be determined by the proportion of male and female staff.

　　3　The building area of staff bathroom shall be determined according to Table 4.2.3.

4.2.4 The building area of one shower in the shower cubicle should be 5.5 m²~6.5 m². The scale of shower cubicle shall be in accordance with Item 1, Article 4.2.3. Auxiliary rooms such as management room shall not be provided for shower cubicle.

Table 4.2.3 **Building Area of Staff Bathroom**

Design capacity (number of persons)	100	200	300	400~600
Building area per shower (m^2)	17~18	11~12	9~10	8~9

Note: The table covers the management room and the water tank room, and does not cover the boiler room.

4.3 Staff Dormitory

4.3.1 The staff dormitory should be centrally planned and constructed for the convenience of management, and shall be located in quiet area with ambient noise complying with relevant standards of the state.

4.3.2 The staff dormitory shall be mainly composed of living room, public activity room and auxiliary room.

4.3.3 The living room should be provided with cable TV and Internet, and shall be provided with restroom. The restroom shall be provided with shower, basin and hot water.

4.3.4 The public rooms should include recreation room, reading room, television room, etc. Auxiliary rooms shall include equipment room and broom closet.

4.3.5 The scale of staff dormitory in the case of new railways shall be determined by the local natural condition and the distribution of cities and towns along the route.

The scale of staff dormitory for railway reconstruction project shall be determined by actual needs.

4.3.6 The design of dormitory shall be in accordance with the current *Code for Design of Dormitory Building* (JGJ 36). The building area shall be 17 m^2~19 m^2 per person, and double room shall be adopted.

4.3.7 When the number of staff is small yet staff dormitory is necessary, dormitory may be arranged in the station (unit).

4.3.8 Every 20 staff members shall be provided with one guest room in the staff dormitory. If there are fewer than 20 staff members, one guest room shall be provided.

4.3.9 The dormitory area should be provided with activity site (sports equipment), centralized greening, bike parking lot, etc.

Appendix A Building Area of Rooms in Railway Workshop, Work Sub-section, EMU Servicing Depot and Work Section

A. 0. 1 The building area of rooms in railway workshop, foreman's section (subdivision) and EMU depot shall be in accordance with Table A. 0. 1.

Table A. 0. 1 Building Area of Rooms in Railway Workshop, Work Sub-section, and EMU Depot

Type of office	Work Sub-section office	Workshop office
Building area (m^2)	140	160~180

Note: The table covers director's office, vice director's office, secretary office, technique room, changing room, driver's room, meeting room and restroom.

A. 0. 2 The building area of rooms in railway work area shall be in accordance with Table A. 0. 2.

Table A. 0. 2 Building Area of Rooms in Work Area

Scale (number of persons)	5	9	14	20	30	40
Building area (m^2)	70~80	80~100	120~140	210~240	310~340	350~380

Notes: 1 The table covers office, duty room, study and meeting room, tools room, materials room, dining room and restroom, and does not cover lounge and dormitory.
  2 Lounges shall be provided in work area. The retiring rooms shall meet the needs of 20% of the staff for work area with dormitory, and shall meet the needs of 50% of the staff for work area without dormitory. The building area of lounge shall be 5~6 m^2 per bed.
  3 When the number of staff members in the work area exceeds 40, the building area shall be determined according to actual needs.

Appendix B Technical Requirements of Communication Equipments Rooms

Computer room name		Technical requirement	Temperature (°C)	Relative Humidity (%)	Minimum indoor height (m) from the bottom surface of beam or ceiling to the ground	Minimum indoor height (m) From the ceiling to the ground when there is no beam	Load on the ground (floor) surface (kN/m²)	Surface layer Material of the ground	Wall surface	Ceiling	Window	Door	Air-conditioning	Ventilation
Communication station		Communication equipment room	18~28	30~75	3.4	3.6	8	Movable anti-static floor is layered on the compacted and light-catching ground surface (the height is 300 mm between the surface and ground)	Dust-proof, light colored, surface electrostatic dissipative performance coating	Small piece of ceiling	Double air tight window	Air tight and thermally insulated door, dust-proof and thermally insulated main door with two doors opened outward, the door height not less than 2.2 m, and the width not less than 1.2 m	Air conditioning dedicated for equipment room operating uninterrupted for 24 h in the whole year shall be installed	—
		Network management room											Air conditioning dedicated for equipment room operating uninterrupted for 24 h in the whole year shall be installed	Ventilation devices are provided
		Monitoring center												—
		Main distribution room	10~32	20~80										—
		Spare parts and instrument	18~28	30~75									Air conditioning dedicated for equipment room operating uninterrupted for 24 h in the whole year shall be installed	—
		Power supply equipment room	16~30	≤75	3.0	3.3	12	Terrazzo	light colored paint is coated, light green wainscot is made under the height of 1.2 m	Paint light colored paint				—

(continued)

Computer room name	Technical requirement	Temperature (°C)	Relative Humidity (%)	Minimum indoor height (m) from the bottom surface of beam or ceiling to the ground	Minimum indoor height (m) From the ceiling to the ground when there is no beam	Load on the ground (floor) surface (kN/m²)	Surface layer Material of the ground	Wall surface	Ceiling	Window	Door	Air-conditioning	Ventilation
Communication station	Battery room	≥14 (If valve control type is used, 18~28)	—	3.0	3.3	15	Acid-fast, 5~8 per thousand drain slope in battery room and electric liquid chamber. The interior design of the pool, separate water pipes(battery with valve control type, the ground is terrazzo) The slope of 0.5%~0.8% shall be made from electric liquid chamber of the battery room to the side of the drainage ditch. Water pool and separate water pipe shall be arranged indoors (when valve control type battery is used, the terrazzo shall be used to make the ground surface	White acid-resistant paint is applied, and 0.3 m acid-resistant cement or ceramic tile base is set. (When the valve control type battery pack is used, the wall surface shall be the same as that of the power equipment room)	Apply white acid-resistant paint (When the valve control type battery pack is used, the ceiling shall be the same as that of the power equipment room)	Double air tight window, acid resistant frosted glass, the area ratio between window and floor is 1/6 to 1/8(when valve control type battery pack is used, double sealed windows shall be used,and the area ratio between window and floor shall be 1/6 to 1/8)	Dust-proof, acid-resistant, outward opening(When the valve control type battery pack is used, dust proof is required and the door shall open outward)	Not installed	Exhaust fan is provided (When valve control type battery pack is used, ventilation is not required)

(continued)

Computer room name	Technical requirement	Temperature (°C)	Relative Humidity (%)	Minimum indoor height(m) from the bottom surface of beam or ceiling to the ground	Minimum indoor height(m) From the ceiling to the ground when there is no beam	Load on the ground (floor) surface (kN/m²)	Surface layer Material of the ground	Wall surface	Ceiling	Window	Door	Air-conditioning	Ventilation
	Lead-in room	≥12	—	—	Same as the rooms of the same floor	8		The same as power equipment room	Common type	Common type	the width of one door not less than 1.0 m	Not installed	
Communication station	Backup room	18~28	30~75	3.4	3.6	8	Terrazzo	Dust-proof, light colored, surface electrostatic dissipative performance coating	Common type	The same as communication equipment room	The same as communication equipment room	Installed in the future	Ventilating fan
Station communications equipment room	Communications equipment room	18~28	30~75	3.4	3.6	8	Movable anti-static floor is layered on the compacted and light-catching ground surface (300 mm from the floor surface to the structural surface)	Dust-proof, light colored, surface electrostatic dissipative performance coating	Common type	Double aluminum alloy air-tight window	Double outward opened doors, the door height is not less than 2.2 m, width is not less than 1.2 m	Air conditioning dedicated for equipment room operating uninterrupted for 24 h in the whole year shall be installed	—
Section communications equipment room		18~28	30~75	3.4	3.6	8	Movable anti-static floor is layered on the compacted and light-catching ground surface (300 mm from the floor surface to the structural surface)	Dust-proof, light colored, surface electrostatic dissipative performance coating	Common type	Not installed	Double outward opened doors, the door height is not less than 2.2 m, width is not less than 1.2 m	Air conditioning dedicated for equipment room operating uninterrupted for 24 h in the whole year shall be installed	—

(continued)

Computer room name	Technical requirement	Temperature (°C)	Relative Humidity (%)	Minimum indoor height (m) from the bottom surface of beam or ceiling to the ground	Minimum indoor height (m) From the ceiling to the ground when there is no beam	Load on the ground (floor) surface (kN/m^2)	Surface layer Material of the ground	Wall surface	Ceiling	Window	Door	Air-conditioning	Ventilation
Section communications equipment room	Section GSM-R repeater room	—	—	3.0	—	—	Terrazzo	Common type	Common type	Common type	Common type	Not installed	—
Tunnel	Equipment chamber in tunnel	—	—	—	—	—	Terrazzo	Common type	Common type	Not installed	The pressure and suction generated by the train operation should be considered for the door	When base station is set, air conditioning dedicated for equipment room operating uninterrupted for 24 h in the whole year shall be installed	

Note: Heating facilities shall be set up in accordance with the geographic location.

Appendix C Technical Requirements for Signaling Equipment Room

Technical requirements / Room type	Temperature (℃)	Relative humidity (%)	Indoor height (m) - From the bottom edge of beam or the bottom ceiling to the ground	Indoor height (m) - From the ceiling to the ground when there is no beam	Load on the ground surface (kN/m²)	Material of the floor surface	Wall surface	Ceiling	Window	Gate	Air-condition	Ventilation
Signaling computer room, unattended relay station in sections (excluding container type room)	15~28	35~75	3.3~3.6; effective height not less than 3.0	3.3~3.6; effective height not less than 3.0	Computer room 8; the location of the UPS is determined according to the calculation of equipment load	Movable anti-static floor is layered on the compacted and light-catching ground surface (from the surface to the ground 300 mm)	Cover paint with dust-free, light-color, static-dissipative surface	Ceiling material shall be fire-proof, anti-flaming and scaling resistant	Windows shall not be provided in unattended signaling equipment room in section, and windows should not be provided in other rooms. Bi-layer airtight windows shall be provided if windows are necessary	Metallic safety gate shall have good air-tightness, the height shall not be less than 2.2 m, and the clear width shall not be less than 1.2 m	Operation without interruption, 24-hour uninterrupted dedicated air conditioner	Installed
Relay room					According to the calculation of equipment load	Tiled ground/anti-static movable floor						
Power supply room		Configured according to equipment maintenance requirement	Determined according to the height of equipment, not less than 5.0 in repairing room	Determined according to the height of the equipment	Configured as required	Painted and tiled ground, repairing room shall be configured according to the onsite requirement						
Maintenance workshop	Not more than 35						Dust-free, light-color, the paint shall have static-dissipative performance in rooms for maintaining electronic equipment	Ceiling material shall be fire-proof, anti-flaming and scaling resistant	Bi-layer airtight windows shall be provided for rooms used for maintaining electronic equipment	Metallic gate whose height and width shall comply with the requirements of manufacturing process	Air condition shall be provided according to the geographic location and separate air conditioner may be adopted	Installed

Note: Heating facilities shall be set according to geographic location.

Appendix D Technical Requirements for Information, Disaster Prevention and Safety Monitoring Equipment Rooms

Technical Requirement / Name of computer Room	Temperature (°C)	Relative humidity (%)	Indoor clear height (m)		Load on the ground or floor (building) (kN/m²)	Material of ground surface	Wall	Ceiling	Window	Door	Air Conditioner	Ventilation
			Under beam or under ceiling	Under ceiling if there is no beam								
Information computer room, public security computer room, disaster prevention and safety monitoring computer room, wiring and device room of information system	18~28	35~75	≥2.6	≥2.6	8	Compacting and light-catching concrete surface with anti-static floor above (300 mm from surface to the ground)	Decoration material with good tightness, no dust, light color, and surface electrostatic dissipation	Grid ceiling	Double airtight window	Airtight and thermally insulated door, dustproof and thermally insulated main door; double doors opening outward, height not less than 2.2 m, and clear width not less than 1.2 m	Dedicated computer room air-conditioner running for 24 h everyday for the whole year	Installed
Integrated supervision room, fire control room	18~28	35~75	≥2.6	≥2.6	8	Compacting and light-catching concrete surface with anti-static floor above (300 mm from surface to the ground)	Decoration material with good tightness, no dust, light color, and surface electrostatic dissipation	Grid ceiling	Double airtight window	Airtight and thermally insulated door, dustproof and thermally insulated main door; double doors opening outward, height not less than 2.2 m, and clear width not less than 1.2 m	Air-conditioner running for 24 h everyday for the whole year	Installed

(continued)

Technical Requirement / Name of computer Room	Temperature (°C)	Relative humidity (%)	Indoor clear height (m)		Load on the ground or floor (building) (kN/m²)	Material of ground surface	Wall	Ceiling	Window	Door	Air Conditioner	Ventilation
			Under beam or under ceiling	Under ceiling if there is no beam								
Power supply room	15~28	35~75	≥3	≥3	according to the loads of device	Ceramic floor	Dust-free decoration material with light color and surface electrostatic dissipation	No requirement on suspended ceiling, ceiling material shall be fireproof, retardant and scaling proof	Double airtight window	Airtight safety door, height not less than 2.2 m, and clear width not less than 1.2 m	Dedicated computer room air-conditioner running for 24 h everyday for the whole year	Installed

Notes: 1 Heating facilities shall be set as required.
2 The indoor clear height is the distance from the ceiling to the ground surface or to the anti-static floor if any.

Appendix E Technical Requirements of Traction Power Supply and Power Distribution Buildings

Category of room	Fire proof requirement	Exit	Clear height and area	Door	Window	Floor and wall
High voltage switchgear and capacitor room	In accordance with the current Code for Design on Fire Prevention of Railway Engineering (TB 10063), Design Code of Railway Electric Traction Feeding (TB 10009), Code for Design of Railway Electric Power (TB 10008) and Code for Design of Fire Protection for Fossil Fuel Power Plants and Substations (GB 50229)	There shall be two exits if the length of switchgear room exceeds 7 m; there should be one additional exit if it exceeds 60 m. For switchgear room not on the ground floor, one of the exits shall directly lead to the outdoor	Determined by the requirements of equipment layout	The doors as exits and doors which have risk of fire and explosion shall be opened outward, and spring lock shall be installed. Doors between Adjacent distribution rooms shall be able to open in both directions. If the doors of oil filled switchgear room are opened to the structure out of the scope of power distribution installations, doors shall be solid and shall be non-flammable or difficult to be inflamed	Windows may be provided but measurements preventing rain, snow and small animals from entering indoors shall be taken	Inner walls adjacent to ceiling and live parts shall be processed, other walls shall be brushed or painted in white, and ground (floor) should be covered with pressed and polished cement. When switchgear equipment is adopted, the flatness and hardness of the ground shall meet the requirements for cabinet installation, and the walls shall be painted
Control room (secondary equipment room)		There should be two exits leading to the outdoor. If control room is not on the ground floor, one of the exits shall lead to the terrace of outdoor	Indoor clear height should be 3.4 m ~ 4.4 m, and may be relatively lowered if air condition is used	Doors shall open outward	There shall be good day light	Removable anti-static floor should be used. Anti-dust material should be adopted for the floors
GIS room		There shall be two exits if the length of switchgear room exceeds 7 m; there should be one additional exit if it exceeds 60 m. For switchgear room not on the ground floor, one of the exits shall directly lead to the outdoor	Determined by actual equipment layout	Doors shall open outward and equipped with spring lock. Doors between adjacent GIS rooms shall be able to open in both directions	Windows may be provided but measurements preventing rain, snow and small animals from entering indoors shall be taken	Indoor floors shall be clean and anti-dust. Wear resistant, anti-skidding, high hardness material should be adopted for floors. Walls shall be painted

(continued)

Category of room	Fire proof requirement	Exit	Clear height and area	Door	Window	Floor and wall
Cable interlayer	In accordance with the current *Code for Design on Fire Prevention of Railway Engineering* (TB 10063), *Design Code of Railway Electric Traction Feeding* (TB 10009), *Code for Design of Railway Electric Power* (TB 10008) and *Code for Design of Fire Protection for Fossil Fuel Power Plants and Sunstations* (GB 50229)	There should be two exits	Dimensions shall be determined by all the cables to be installed. Cable arrangement shall not disturb safe operation. The requirements of cable laying operation and inspection shall be met. Floor height should be 2.3 m~2.6 m. Distance between beam bottom and the ground floor shall not be less than 2 m	—	—	Ground and walls shall be brushed in white or painted. Ground (floor) should be covered with cement, of which the strenth shall meet the requirements for cable support fastening
Transformer room		There shall be two exits if the length of transformer room exceeds 7 m	Determined by actual equipment layout	Doors shall open outward and be equipped with spring lock. If the doors open to the structure out of the scope of power distribution installations, they shall be solid and shall be nonflammable or difficult to be inflamed	—	Wear resistant, anti-skidding, high hardness material should be adopted for indoor floors
Dispatching room in dispatching center and computer room for SCADA system		There should be two exits leading to the outdoor	Ceiling should be arranged for dispatching room, clear height from the ground floor to the ceiling should not be less than 3.5 m	Doors shall open outward	—	Wall surface shall be applied with dust-free, light-color, static dissipative painting. Ground (floor) should be of pressed and polished cement, with removable anti-static floor above
Power supply room in dispatching center		There shall be two exits if the length of power supply room exceeds 7 m	Determined by actual equipment layout	Doors shall open outward and be equipped with spring lock	—	Wall surfaces shall be brushed and painted in white. Ground (floor) should be covered with polished cement, which shall be pressed

Explanation of Wording in this Standard

Words used for different degrees of strictness are explained as follows in order to mark the differences in executing the requirements in this Standard.

(1) Words denoting a very strict or mandatory requirement:

"Must" is used for affirmation;

"Must not" for negation.

(2) Words denoting a strict requirement under normal conditions:

"Shall" is used for affirmation;

"Shall not" for negation.

(3) Words denoting a permission of a slight choice or an indication of the most suitable choice when conditions permit:

"Should" is used for affirmation;

"Should not" for negation.

(4) "May" is used to express the option available, sometimes with the conditional permit.

Citations

GBZ 1 *Hygienic Standards for the Design of Industrial Enterprises*
GBJ 87 *Code for Design of Noise Control of Industrial Enterprises*
GB 146.2 *Structure Gauge for Standard Gauge Railways*
GB 50016 *Code for Fire Protection Design of Buildings*
GB 50046 *Code for Anticorrosion Design of Industrial Constructions*
GB 50074 *Code for Design of Oil Depot*
GB 50156 *Code for Design and Construction of Filling Station*
GB 50174 *Code for Design of Electronic Information System Room*
GB 50223 *Standard for Classification of Seismic Protection of Building Constructions*
GB/T 50226 *Code for Design of Railway Passenger Station Buildings*
TB 10004 *Code for Design of Railway Locomotive Facilities*
TB 10007 *Code for Design of Railway Signaling*
TB 10009 *Design Code of Railway Electric Traction Feeding*
TB 10010 *Code for Design of Water Supply and Sewerage of Railway*
TB 10029 *Code for Design of Railway Passenger Car Rolling Stock Facilities*
TB 10031 *Code for Design of Railway Freight Car Rolling Stock Facilities*
TB 10057 *Design Specification for Running Safety Monitoring System of Rolling Stock*
TB 10062 *Code for Design on Hump and Marshalling Yard of Railway*
TB 10063 *Code for Design on Fire Prevention of Railway Engineering*
TB 10067 *Code for Design of Passenger & Freight Equipment for Railway Station and Yard*
JGJ 36 *Code for Design of Dormitory Building*
JGJ 64 *Code for Design of Restaurant Building*
JGJ 67 *Design Code for Office Building*
YD/T 5003 *Specifications of Engineering Design for Telecommunication Private Premise*
YD 5054 *Classification Standard for Seismic Protection of Telecommunication Buildings*